WATCHING THE WEATHER

When It's
STORMY

Written by
Noah Leatherland

All rights reserved.
Printed in India.

A catalogue record for this book is available from the British Library.

ISBN: 978-1-80505-599-0

Written by:
Noah Leatherland

Edited by:
Elise Carraway

Designed by:
Jasmine Pointer

©2024
BookLife Publishing Ltd.
King's Lynn, Norfolk
PE30 4LS, UK

FSC MIX
Paper | Supporting responsible forestry
FSC® C195953

All facts, statistics, web addresses and URLs in this book were verified as valid and accurate at time of writing. No responsibility for any changes to external websites or references can be accepted by either the author or publisher.

AN INTRODUCTION TO BOOKLIFE RAPID READERS...

Packed full of gripping topics and twisted tales, BookLife Rapid Readers are perfect for older children looking to propel their reading up to top speed. With three levels based on our planet's fastest animals, children will be able to find the perfect point from which to accelerate their reading journey. From the spooky to the silly, these roaring reads will turn every child at every reading level into a prolific page-turner!

CHEETAH
The fastest animals on land, cheetahs will be taking their first strides as they race to top speed.

MARLIN
The fastest animals under water, marlins will be blasting through their journey.

FALCON
The fastest animals in the air, falcons will be flying at top speed as they tear through the skies.

Photo Credits
Images are courtesy of Shutterstock.com. With thanks to Getty Images, Thinkstock Photo and iStockphoto.
Cover – MeSamong, ilnazgilov, Artistic Design 24, Technicsorn Stocker. Texture throughout – MeSamong. 4–5 – Torychemistry, justkgoomm, Gwens Graphic Studio. 6–7 – Pictureguy, Flystock, Natalllenka.m. 8–9 – Alexpopov, Roman Mikhailiuk. 10–11 – Timodaddy, Pau Buera, Dark Moon Pictures. 12–13 – Fernando Astasio Avila, Nikolay Zaborskikh, Anatolir. 14–15 – Sezamnet, New Africa, HappyPictures. 16–17 – Minerva Studio, Todd Shoemake. 18–19 – Ryan DeBerardinis, Artsiom P, 1Arts. 20–21 – Joko P, Menno van der Haven, robuart. 22–23 – George Trumpeter, Cammie Czuchnicki.

CONTENTS

PAGE 4	What Is Weather?
PAGE 6	What Is a Storm?
PAGE 8	Thunderstorms
PAGE 10	Lightning
PAGE 12	Thunder
PAGE 14	Helpful Storms
PAGE 16	Tornadoes
PAGE 18	Hurricanes
PAGE 20	Staying Safe
PAGE 22	Stormy Days
PAGE 24	Glossary and Index

WORDS THAT LOOK LIKE THIS ARE EXPLAINED IN THE GLOSSARY ON PAGE 24.

What Is WEATHER?

The weather is what it is like in the sky around you. The weather is always changing around the world.

There are different types of weather. It could be hot, cold, wet or dry. It could be a mixture.

What Is a STORM?

Storms are a type of **EXTREME** weather. They usually bring lots of heavy rain, strong winds, thunder and lightning.

There are different types of storms. A storm in one place can be very different to a storm somewhere else.

THUNDERSTORMS

Thunderstorms start when it is warm and **HUMID**. Water droplets in the warm air rise, and clouds start to form.

Water droplets in the clouds freeze into tiny pieces of ice. The **ENERGY** from the ice bumping together can cause lightning.

LIGHTNING

Sometimes, lightning stays in the clouds.

Other times, lightning comes out of the clouds and hits the ground.

Lightning makes the air around it very hot, very quickly. This fast heating of air creates the sound of thunder.

THUNDER

In a storm, you see lightning before you hear thunder. This is because light moves much faster than sound.

Count the seconds between lightning and thunder. The longer it takes to hear the thunder, the farther away the storm is.

HELPFUL STORMS

Storms are very important for the environment. Thunderstorms usually happen after a stretch of warm weather.

Thunderstorms often bring lots of rain. Rain helps cool things down and gives water to the plants that need it.

TORNADOES

Storms bring strong winds with them. If the winds are blowing a certain way, they can create a tornado.

Tornadoes are like tubes of spinning air. They stretch from the clouds to the ground. They can be very dangerous.

HURRICANES

Hurricanes have winds that blow in a circle. Hurricanes start over warm ocean waters, but they can reach the land.

The middle of a hurricane is called the eye. It is calm inside the eye, but very windy around it.

STAYING SAFE

Storms are not the most pleasant kind of weather. Scientists can often **PREDICT** when storms are coming and warn you.

It can be dangerous to be outside in a bad storm. Stay away from trees. Lightning often hits tall objects.

STORMY DAYS

Storms can be scary, but they do not have to ruin your day. There is still plenty you can do indoors.

Storms do not usually last for very long. There might even be a rainbow after the storm passes!

GLOSSARY

ENERGY — a type of power, such as light or heat, that can be used to do something

EXTREME — much beyond what is normal or expected

HUMID — when the air contains a high level of water

PREDICT — to make a guess about something that could happen based on evidence and facts

INDEX

CLOUDS 8–10, 17
ICE 9
LIGHTNING 6, 9–13, 21
OCEANS 18
RAIN 6, 15, 23
THUNDER 6, 8, 11–15
TREES 21
WIND 6, 16, 18